EXPERIENCING THE BEAUTY OF MATHEMATICS THROUGH QUOTATIONS

BY THEONI PAPPAS

— WIDE WORLD PUBLISHING/TETRA —

1st Printing September 1995

Wide World Publishing/Tetra
P.O. Box 476
San Carlos, CA 94070

Printed in the United States of America

ISBN:1-884550-04-5

Library of Congress Cataloging–in–Publication Data
Pappas, Theoni.
The music of reason : experiencing the beauty of mathematics through quotations / Theoni Pappas.
p. cm.
ISBN: 1-884550-04-5 (pbk.)
1. Mathematics - - Quotations, maxims, etc. I. Title.
QA99. p38 1995
510 - - dc20 95-38362
CIP

Contents

Mathematics

Mention the word mathematics and one rarely gets an indifferent response. Rather, the emotions range from fear and hate to love and delight. Some people boast with pride of their ignorance of mathematics. Others go to almost any means to avoid it. While a few become engrossed to the point of obsession.

What's so unusual about a subject that evokes such strong and diverse feelings?

Mathematics is a figment of the imagination. All its elements, objects, axioms, theorems, definitions ... describe objects which do not actually exist in our world. The worlds created by mathematicians are imaginary. They exist independently of anything in our world. Granted mathematics can be used to describe, explain and predict phenomena of the universe; but mathematical objects were not necessarily created with that purpose in mind.

What stimulates the mathematician's imagination? Perhaps an interesting pattern, an intriguing problem, a theorem or postulate, a challenge to determine a solution, an unanswered question, or simply curiosity and determination to discover a truth.

Over and over mathematical ideas conceived centuries earlier somehow find their nitch or use in the future. Coincidental?

This, therefore, is mathematics: she reminds you of the invisible form of the soul; she gives to her own discoveries; she awakens the mind and purifies the intellect; she brings light to our intrinsic ideas; she abolishes oblivion and ignorance which are ours by birth.

—Proclus
(c. AD 410-485)

The study of mathematics is apt to commence in disappointment....we are told that by its aid the stars are weighed and the billions of molecules in a drop of water are counted. Yet, like the ghost of Hamlet's father, this great science eludes the efforts of our mental weapons to grasp it.

—Alfred North Whitehead
(1861-1947)

Mathematics may be defined as the subject in which we never know what we are talking about, nor whether what we are saying is true.

—Bertrand Russell
(1872-1970)

$\{\ldots -3, -2, -1, 0, 1, 2, 3 \ldots\}$

$\sqrt{2}, \sqrt{3}, \sqrt{5}, \sqrt{7}, \ldots$

$\{\} = \phi$

$\sqrt{-1} = i$

$\frac{\bigcirc}{d} = \pi$

I am one and two,
1/2 and 29/30,
0.03 and 6.3333,
-8 and e,
a million and a
googol,
7 and π*;*

I am i and 5+3i;

I am nothing and
zero;

I am the set of all
numbers;
I am the empty set;

The essence of mathematics is its freedom.

— **Georg Cantor**
(1845-1918)

I can rejoice over this perfection and bear witness to it with a clear conscience, for it was not I who invented it or even discovered it. The laws of mathematics are not merely human inventions or creations. They simply are; they exist quite independently of the human intellect. The most that any ...with a keen intellect can do is to find out that they are there and to take cognizance of them.

—**M. C. Escher**
(1898-1972)

It may well be doubted whether, in all the range of science, there is any field so fascinating to the explorer—so rich in hidden treasures—so fruitful in delightful surprises—as Pure Mathematics.

—**Charles Dodgson**
(Lewis Carroll 1832-1898)

$5 + 3 = 8$

$17 - 9 = 8$

I am adding and
subtracting,
multiplying and
dividing;

$23 \times 7 = 161$

$3x^2 - 7x + 8 = 0$

I am a quadratic
equation,
a polynomial,
a coefficient,
a power and
exponent;

I am squares, fractals…
I am a point,
a line, a plane;
I am space,

The beauty in mathematics is seeing the truth without effort.

—George Polya
(1887-1985)

...there is no more a math mind, than there is a history or an English mind...

—Gloria Steinem
(1935-)
Moving Beyond Words

In mathematics there are no true controversies.

—Karl Friedrich Gauss
(1777-1855)

When we cannot use the compass of mathematics or the torch of experience ... it is certain we cannot take a single step forward.

—Voltaire
(1694-1778)

The charm (of mathematics) lies chiefly...in the absolute certainty of its results; for that is what, beyond all mental treasures, the human intellect craves for. Let us be sure of something! More light, more light!

—Charles Dodgson
(Lewis Carroll 1832-1898)

Mathematics consists of proving the most obvious thing in the least obvious way.

—Henri Poincaré
(1854-1912)

Mathematics seems to endow one with something like a new sense.

—Charles Darwin
(1809-1892)

Mathematics takes us into the region of absolute necessity, to which not only the actual world, but every possible world, must conform.

—Bertrand Russell
(1872-1970)

I am the pattern on
a tortoise shell,
a spider's web,
the shape of a leaf;

I am the sound of music,
the crest of a wave;

In mathematics I can report no deficiency, except it be that men do not sufficiently understand the excellent use of Pure Mathematics.

—Francis Bacon
(1561-1626)

Mathematics is the gate and key of the sciences.... Neglect of mathematics works injury to all knowledge, since one who is ignorant of it cannot know the other sciences or the things of this world. And what is worst, those who are thus ignorant are unable to perceive their own ignorance and so do not seek a remedy.

— Roger Bacon
(1214-1294)

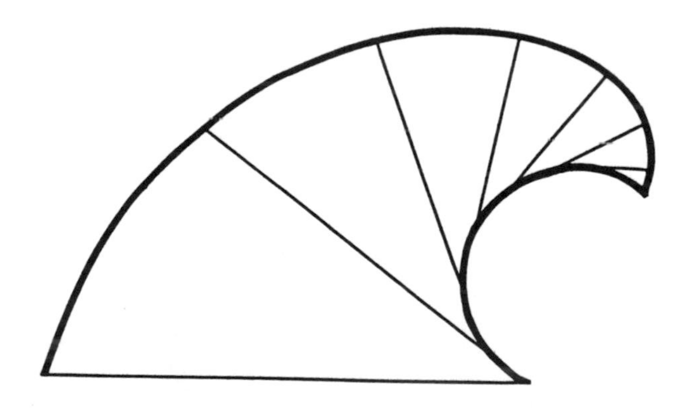

I am the curve of a shark's fin,
the arc of a pendulum
and a unit of time;

Beside the mathematical arts there is no infallible knowledge except it be borrowed from them.

—Robert Recorde
(1510-1558)

A traveler who refuses to pass over a bridge until he has personally tested the soundness of every part of it is not likely to go far; something must be risked, even in mathematics.

— Horace Lamb
(1913-)

If a man's wit be wandering, let him study mathematics.

— Francis Bacon
(1561-1626)

Do not imagine that mathematics is hard and repulsive to common sense.

— Sir William Thomson, Lord Kelvin
(1824-1907)

The true spirit of delight…is to be found in mathematics as surely as in poetry.

— Bertrand Russell
(1872-1970)

Mathematics is thought moving in the sphere of complete abstraction from any particular instance of what it is talking about.

— Alfred North Whitehead
(1862-1947)

I am unreal worlds;

I am order;

I am chaos;

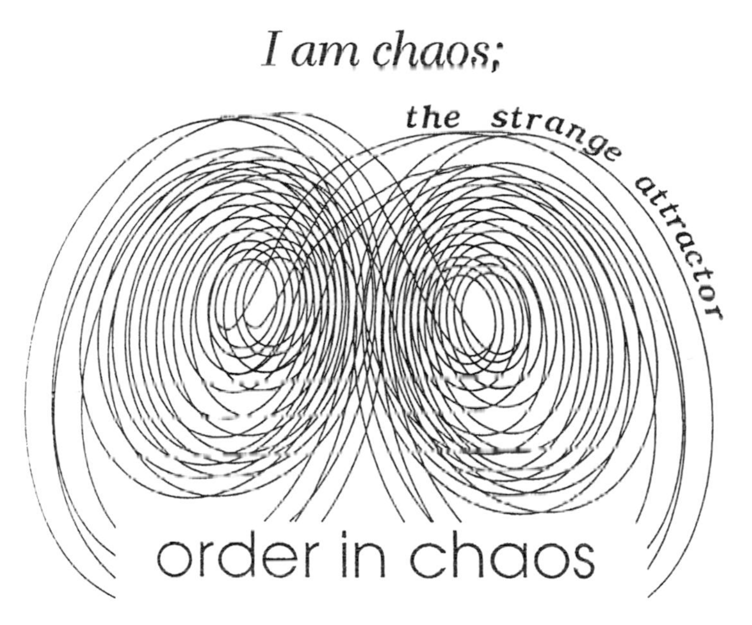

I am MATHEMATICS.

— Theoni Pappas
(1944-)

Mathematics & the Imagination

Imagination is the creative force of an individual. It is the means by which we tackle problems — be they everyday problems or crucial global problems. Without imagination new solutions are difficult to come by. Without imagination a writer cannot create, a composer must give up music and an artist can only paint what he or she can see. Without imagination, mathematics would not exist.

From the creation of the first elements of mathematics — the natural numbers — mathematicians created fiction. Most people don't consider numbers a figment of the imagination because we immediately conceptualize a quantity when we hear a number such as "five". Many different symbols have been invented to stand for "five" — but regardless of the symbols used the quantity conjured up in the mind is the same for everyone.

Is mathematics the framework of our universe(s)?

Is the mathematician merely discovering things which are out there, but have not as yet been discovered? Things that guide the workings of the world?

Some mathematics has been created out of necessity to solve practical everyday problems. But more often than not, mathematical ideas arise as the mathematician's imagination and curiosity wander in their search to understand how mathematical things work.

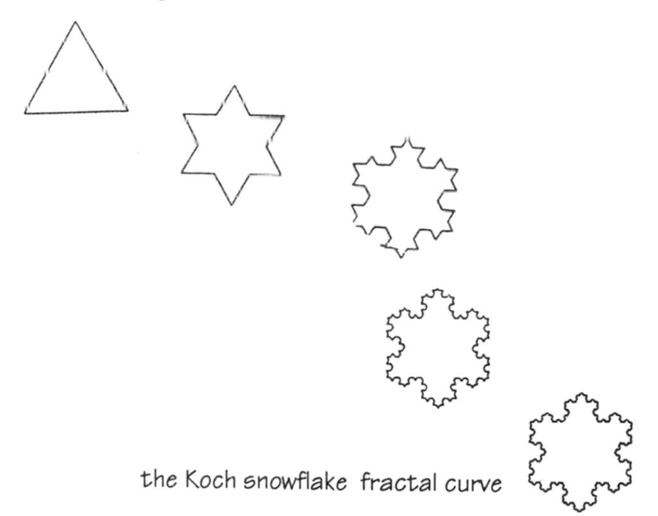

the Koch snowflake fractal curve

There is an astonishing imagination even in the science of mathematics.

—Voltaire
(1694-1778)

The moving power of mathematical invention is not reasoning but imagination.

—Augustus de Morgan
(1806-1871)

The science of pure mathematics, in its modern developments, may claim to be the most original creation of the human spirit.

—Alfred N. Whitehead
(1861-1947)

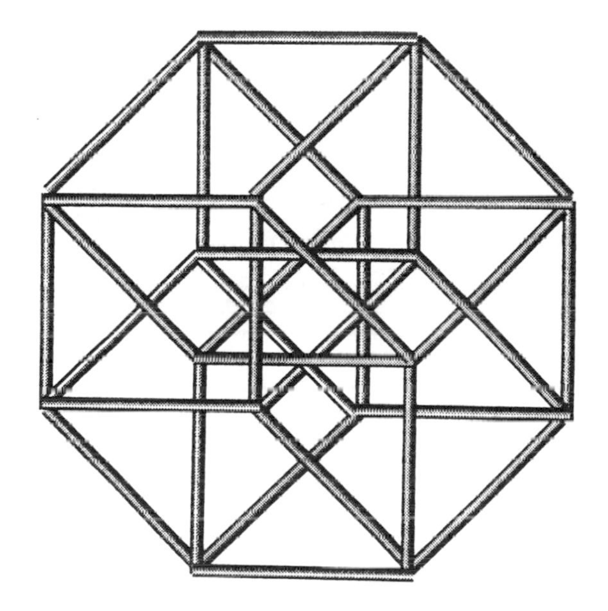

A rendition of a hypercube
(a 4th-dimensional cube)

Many who have never had an opportunity of knowing any more about mathematics confuse it with arithmetic, and consider it an arid science. In reality, however, it is a science which requires a great amount of imagination.

—Sonya Kovalevsky
(1850-1891)

'Mathematizing' may well be a creative activity of man, like language or music, of primary originality...

—Herman Weyl
(1885-1955)

Why, sometimes I've believed as many as six impossible things before breakfast.

— Lewis Carroll
(Charles Dodgson 1832-1898)

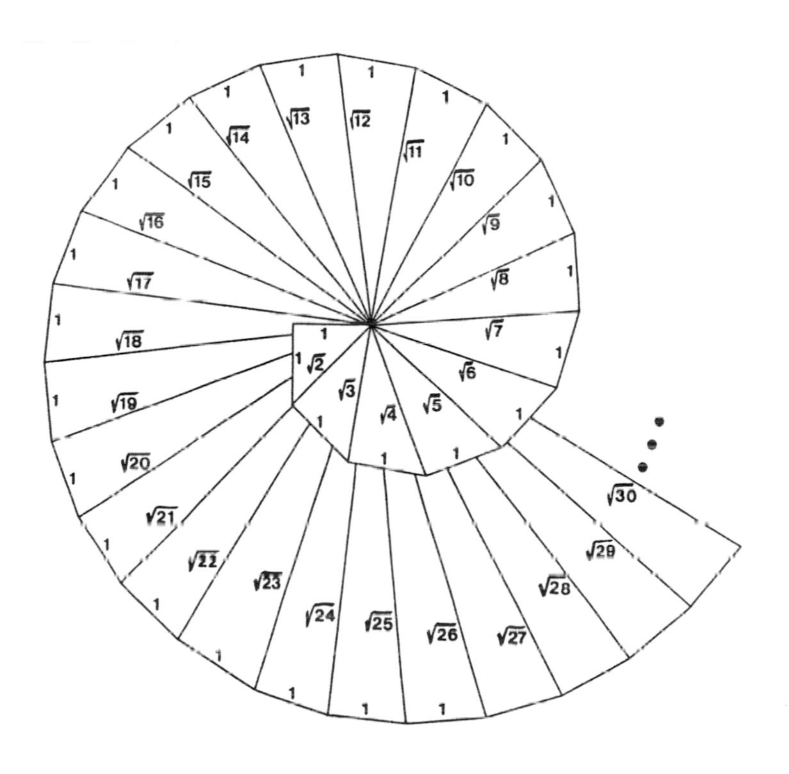
1
1
√2
√3
√4
√5
√6
√7
√8
√9
√10
√11
√12
√13
√14
√15
√16
√17
√18
√19
√20
√21
√22
√23
√24
√25
√26
√27
√28
√29
√30

Imagination is more important than knowledge.

—Albert Einstein
(1879-1955)

The creative act owes little to logic or reason...The creative process cannot be summoned at will or even cajoled by artificial offering. Indeed it seems to occur most readily when the mind is relaxed and the imagination roaming freely.

—Morris Kline
(1908-1992)

Though the source be obscure, still the stream flows on.

—Henri Poincaré
(1854-1912)

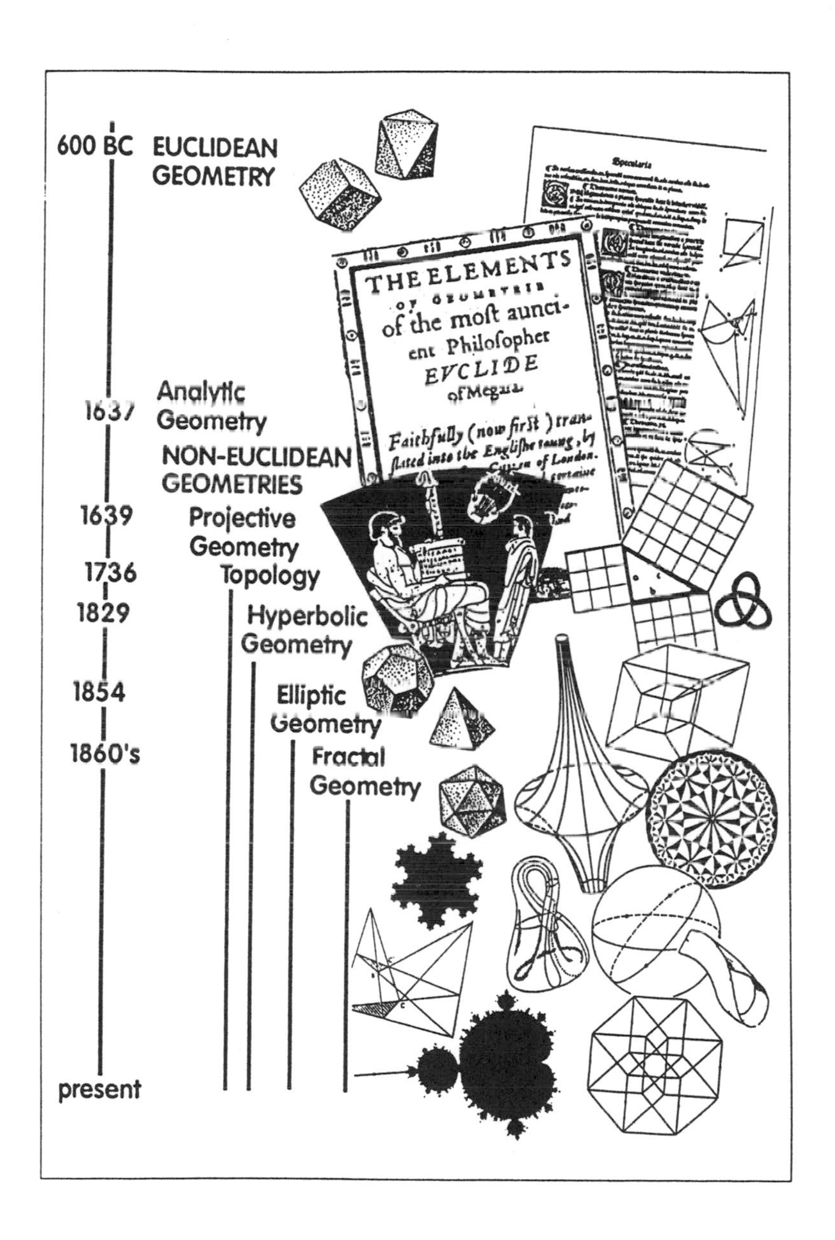
600 BC
EUCLIDEAN GEOMETRY
1637
Analytic Geometry
NON-EUCLIDEAN GEOMETRIES
1639
Projective Geometry
1736
Topology
1829
Hyperbolic Geometry
1854
Elliptic Geometry
1860's
Fractal Geometry
present
THE ELEMENTS
OF GEOMETRIE
of the most aunci-
ent Philosopher
EVCLIDE
of Megara
Faithfully (now first) tran-
slated into the Englishe toung, by
Specularia

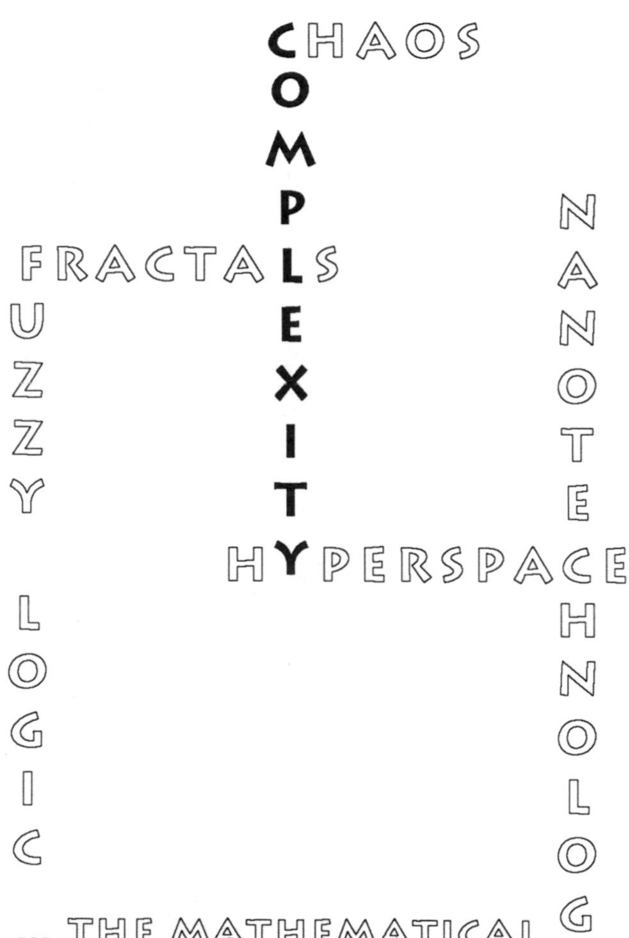
CHAOS
COMPLEXITY
FRACTALS
FUZZY LOGIC
NANOTECHNOLOGY
HYPERSPACE
... THE MATHEMATICAL MIND AT WORK TODAY

...it seems to be true that many things have...an epoch in which they are discovered in several places simultaneously, just as the violets appear on all sides in the springtime.

—Wolfgang Bolyai
(1775-1856)

...if a number is merely the product of our mind, space has a reality outside our mind whose laws we cannot a priori completely prescribe.

—Karl Friedrich Gauss
(1777-1855)

...what the imagination seizes as beauty must be truth— whether it existed before or not.

—John Keats
(1795-1821)

Mathematics & Humor

Mathematics is no laughing matter, or is it? Mathematics is a serious subject. In fact, many feel it is too serious, too sterile, too complex, too dry for a casual conversation. Yet most agree it demands respect. If you've ever watched a math enthusiast involved with a mathematical challenge — the intensity, the pleasure and emotions expressed strike a non-math person as peculiar or alien. So what can be so funny about math?

Mathematics is fertile ground for humor. It creates difficulties and frustrations for some and enjoyment for others; it has humiliated and caused anxiety for some while making others feel elated. These contrasting emotions make it an ideal topic with which to poke fun. Without being able to laugh at our shortcomings, addictions and pleasures, how would we be able to admit they exist, face them, come to grips with them and perhaps resolve them?

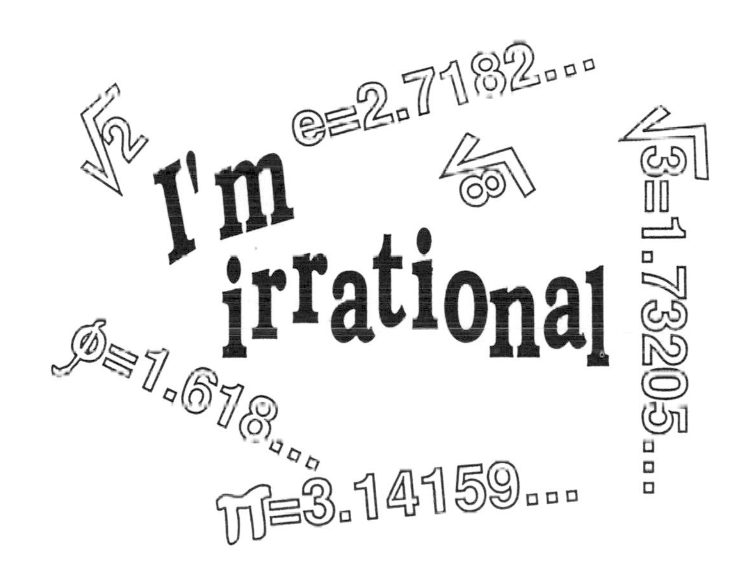
√2
e=2.7182...
I'm irrational
√8
√3=1.73205...
Φ=1.618...
π=3.14159...

One cannot escape the feeling that these mathematical formulae have an independent existence and an intelligence of their own, that they are wiser than we are, wiser even that their discoverers, that we get more out of them than was originally put into them.

—Heinrich Hertz
(1857-1894)

I have hardly known a mathematician capable of reasoning.

—Plato
(427?-347 B.C.)

I'm no mathematician. I find counting a bore.
—Madonna as Breathless
in the movie *Dick Tracy,* 1990

The different branches of Arithmetic-Ambition, Distraction, Uglification, and Derision.

—Lewis Carroll
(Charles Dodgson 1832-1898)
Alice in Wonderland

I had been to school...and could say the multiplication tables up to 6 x 7 = 35, and don't reckon I could ever get any further than if I was to live forever. I don't take stock of mathematics.

—Mark Twain
(Samuel Langhorne Clemens 1835-1910)
Huckleberry Finn

Oh these mathematicians make me tired! When you ask them to work out a sum they take a piece of paper, cover it with rows of A's B's and X's Y's...scatter a mess of flyspecks over them, and then give you an answer that's all wrong.

—Thomas A. Edison
(1847-1931)

Do not worry about your difficulties in mathematics. I can assure you that mine are still greater.

—Albert Einstein
(1879-1955)

Mathematics is the art of giving the same name to different things.

— Henri Poincaré
(1854-1912)

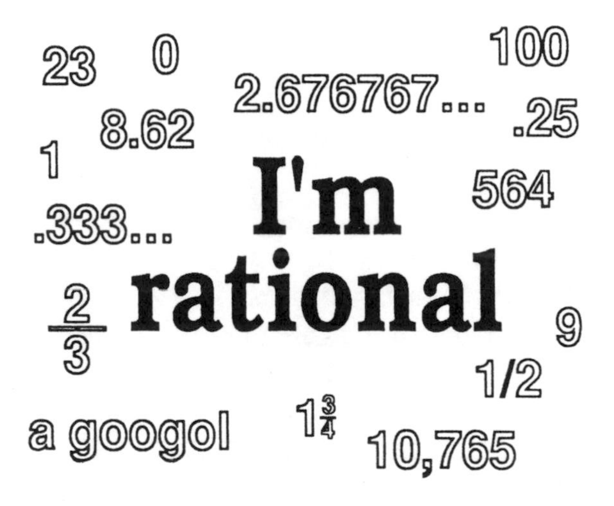

Math is like learning a foreign language, Marcie. No matter what you say, it's going to be wrong anyway!

— **Peppermint Patty**
in *Peanuts* by
Charles Schulz
(1922-)

Science is always wrong; it never solves a problem without creating ten more.

— **George Bernard Shaw**
(1856-1950)

Algebra begins with the unknown and ends with the unknowable.

— **Anonymous**

A man has one hundred dollars and you leave him with two hundred dollars, that's subtraction.

— **Mae West**
in *My Little Chickadee*, 1940

...Mathematics may be defined as the subject in which we never know what we are talking about, nor whether what we are saying is true.

— **Bertrand Russell**
(1872-1970)

Mathematicians

Who is the mathematician? One who deals with abstractions, creates imaginary worlds complete with all the characters interacting in specific ways. The mathematician is not satisfied with just solving a problem, but often wants to exhaust all possible ways of solving it. One proof of the Pythagorean theorem would suffice for most people, but not the mathematician. One geometry is enough for most of us to study, but not for mathematicians. Give them a problem to solve, and they are in heaven. Mathematicians rise to the challenge of mental marathons. What other group would one find spending centuries trying to prove Fermat's Last Theorem, let alone all the other "impossible" problems that have been solved or are awaiting a new mathematician to tackle. Who looks for ever larger prime numbers? ever more accurate decimal representation of π? twin primes? Who solves a pastime —Königsberg bridges walk — and launches a new mathematical field?

Here is what mathematicians and others have to say about this unusual breed, and here are some of the famous problems they've tackled and some they are still tackling.

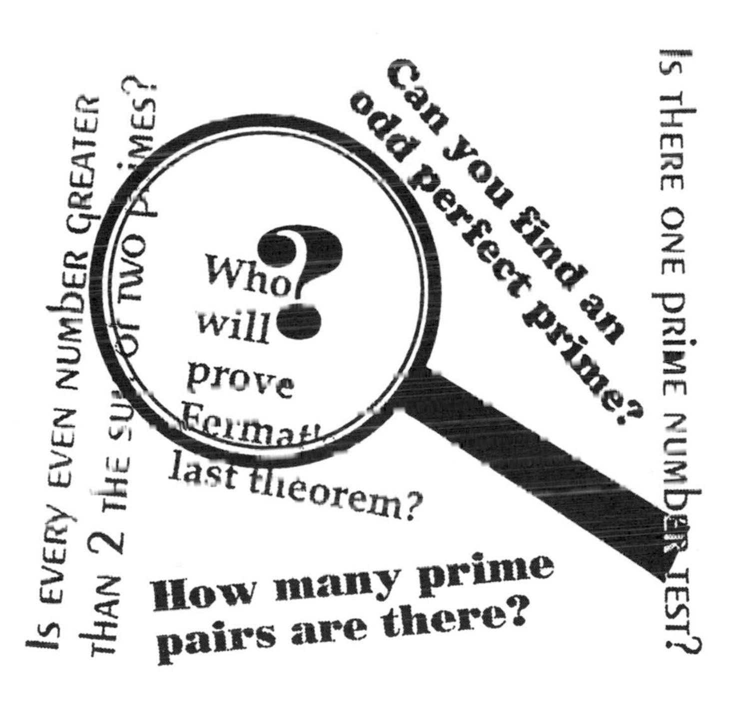

The ancient
impossible trio—

you can't

• trisect an angle
• duplicate a cube
• square a circle

with just a
compass
and a
straightedge?

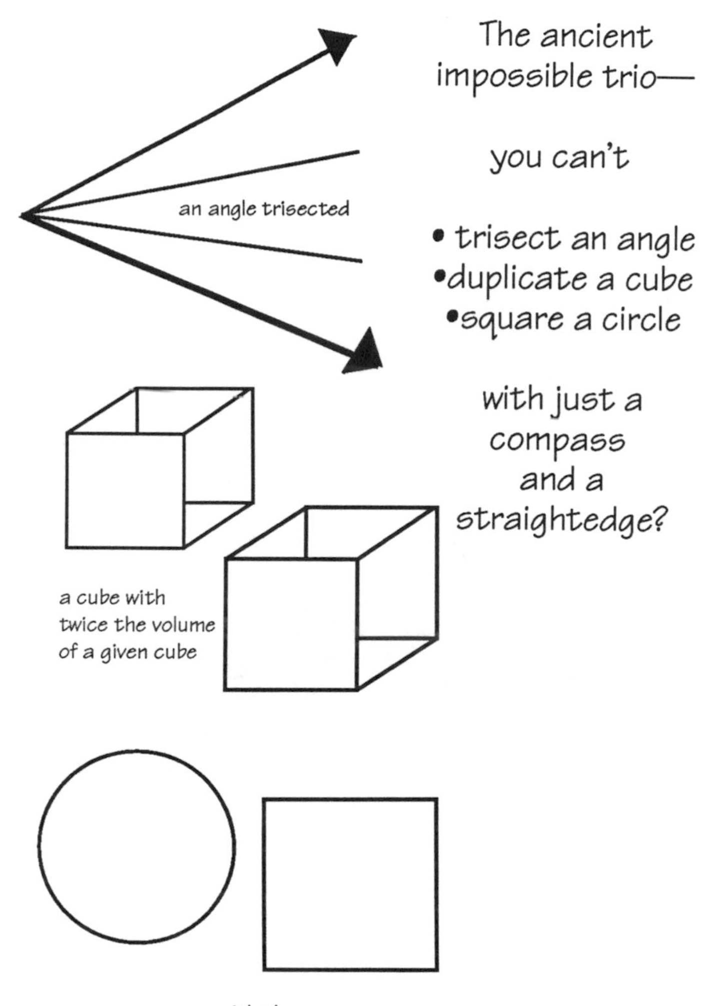

a square with the same
area of a given circle

There is no sharply drawn line between those contradictions which occur in the daily work of every mathematician... and the major paradoxes which provide food for logical thought for decades and sometimes centuries.

— Nicholas Bourbaki
(collective name for a predominately French group of mathematicians, since 1939)

The mathematician does not study pure mathematics because it is useful; he studies it because he delights in it and he delights in it because it is beautiful.

— Henri Poincaré
(1854-1912)

Mathematicians are species of Frenchmen; if you say something to them, they translate it into their own language and presto! it is something entirely different.

—Johann Wolfgang von Goethe
(1749-1832)

1
1 1
1 2 1
1 3 3 1

Pascal probability Laplace &

Chu Shin–Chieh

Zeno paradox Whitehead

Liber Abaci Fibonacci sequence

1, 1, 2, 3, 5 ,8, . . .

Pythagoras $a^2 + b^2 = c^2$

$leg^2 + leg^2 = hypotenuse^2$

al Khowarizmi algebra Bhaskara

Euclid geometries Riemann &
Bolyai & Lobachesky

set theory Cantor
transfinite numbers

Babbage computers
Lovelace programming

The pure mathematician who should forget the existence of the exterior world would be like a painter who knows how to harmoniously combine colors and forms, but who lacked models. His creative powers would soon be exhausted.

—Henri Poincaré
(1854-1912)

A mathematician, like a painter or poet, is a maker of patterns...with ideas.

—Godfrey H. Hardy
(1877-1947)

The mathematician may be compared to a designer of garments, who is utterly oblivious of the creatures whom his garments may fit.

—Tobias Danzig
(1884-1956)

Let us grant that the pursuit of mathematics is divine madness of the human spirit, a refuge from the goading urgency of contingent happenings.

—Alfred North Whitehead
(1861-1947)

I hope that posterity will judge me kindly, not only as to the things which I have explained, but also to those which I have intentionally omitted so as to leave to others the pleasure of discovery.

—René Descartes
(1596-1650)

The mathematician has reached the highest rung on the ladder of human thought.

—Havelock Ellis
(1859-1939)

I have hardly known a mathematician capable of reasoning.

—Plato
(427?-327? B.C.)

Where the statue stood
Of Newton, with his prism
and silent face,
The marble index of a mind forever
Voyaging through
strange seas of thought, alone.

—William Wordsworth
(1770-1850) *The Prelude*

Einstein relativity

Loyd puzzles Dudeney

Napier logarithms

Apollonius conics Hypatia

Agnesi curves & analysis Noether

Descartes Cartesian coordinates

calculator Pascal & Leibnitz

Newton calculus Leibnitz & Seki Kowa

Diophantus equations

Eratosthenes earth measurement

...it is impossible to be a mathematician without being a poet in soul....imagination and invention are identical...the poet has only to perceive that which others do not perceive, to look deeper than others look. And the mathematican must do the same thing.

—Sonya Kovalevsky
(1850-1891)

Mathematics & the Arts

Mathematics is a science, a language, a way of thinking. Above all, it is an art. The mathematician paints with numbers, equations, and abstract objects. Yet the mathematical palette also helps enhance the work of artists, musicians, architects, sculptors. Mathematical concepts — such as the golden rectangle, tessellations, projective geometry, symmetry, perspective, geometries, higher dimensions, infinity, space, center of gravity, knots, topology, solids, computer science ... —influence the arts. From the ancient Greek sculptor, Phidias, to modern artists such as Picasso, Escher, Dali, Tony Robbin — mathematical ideas have played invaluable roles in their works. The ancients used mathematical ratios to explore musical sounds. Today wavelets, Fourier series, computer modeling digitization, and superstrings continue to explain and delve into the mathematics of music.

May not Music be described as the Mathematics of sense, Mathematics as the Music of the reason?

— James Joseph Sylvester

(1814-1897)

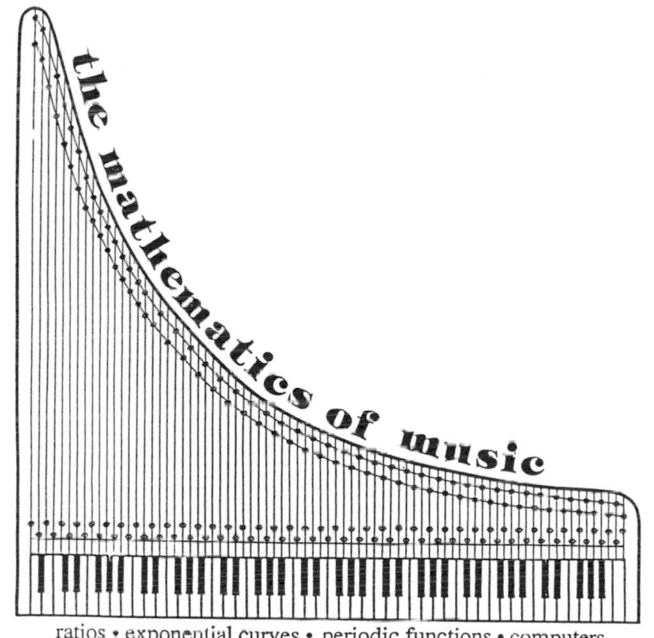

Music is the pleasure the human soul experiences from counting without being aware it is counting.

—Gottfried Leibniz

(1646-1716)

Drawing by Leonardo da Vinci.

Mathematics rightly viewed possesses not only truth, but supreme beauty—a beauty cold and austere, like that of sculpture. ..yet sublimely pure, and capable of a stern perfection such as only the greatest art can show.

— Bertrand Russell
(1872-1970)

There is music even in the beauty, and the silent note which Cupid strikes, far sweeter than the sound of an instrument; for there is music wherever there is harmony, order, or proportion; and thus far we may maintain the music of the spheres.

—Sir Thomas Browne
(1605-1683)

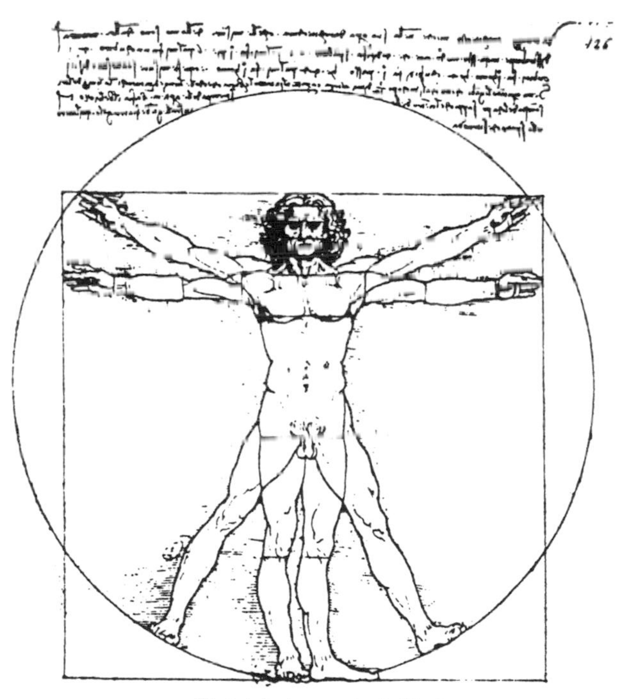

Sketch by Leonardo da Vinci.

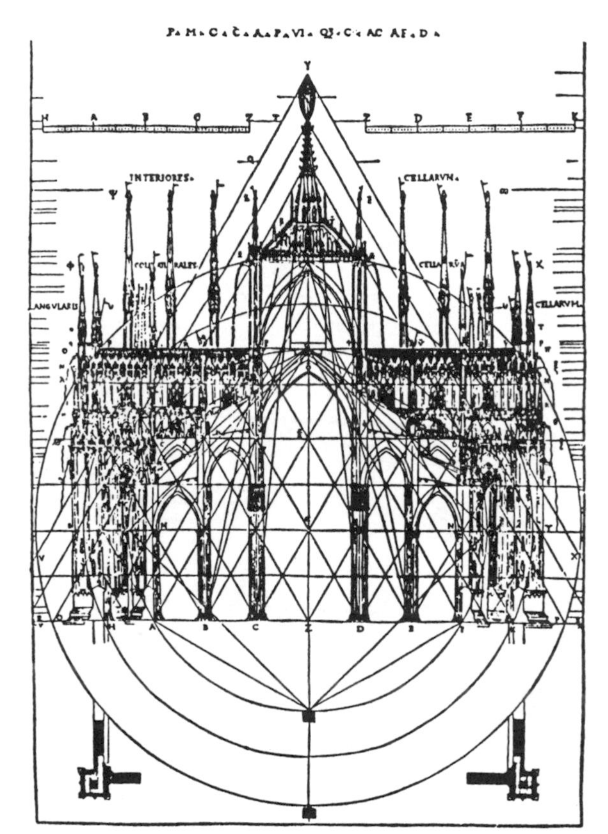

Gothic architectural plans published in 1521 by master architect of the Dome of Milan, Caesar Caesariano.

Architecture is akin to music in that both should be based on the symmetry of mathematics.

—Frank Lloyd Wright
(1867-1959)

...a good part of art is based on geometry...

—Giraud Desargues

(1591-1661)

Might is geometry; joined with art, resistless.

— Euripides

(480?-406 B.C.)

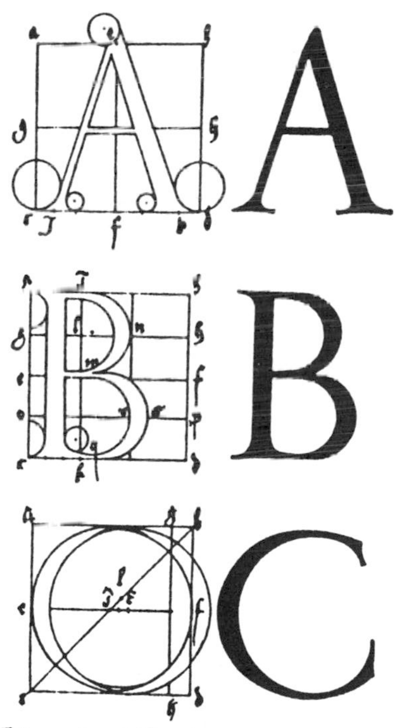

Typography of Albrecht Dürer (1471-1528)

The unfolded hypercube was the inspiration
for Salvador Dali's The Crucifixtion (1956).

The most beautiful thing we can experience is the mysterious. It is the source of all true art and science. He to whom this emotion is a stranger, who can no longer pause to wonder and stand wrapped in awe, is as good as dead.

—Albert Einstein
(1879-1955)

Here, where we reach the sphere of mathematics, we are among processes which seem to some the most inhuman of all human activities and the most remote from poetry. Yet it is here that the artist has the fullest scope of his imagination.

—Havelock Ellis
(1859-1939)

...the feeling of mathematical beauty, of the harmony of numbers and of forms, of geometric elegance. It is a genuinely esthetic feeling, which all mathematicians know. And this is sensitivity.

—Henri Poincaré
(1854-1912)

Mathematics & Infinity

Infinity has stimulated imaginations for thousands of years. It is an idea drawn upon by theologians, poets, artists, philosophers, writers, scientists, mathematicians — an idea that has perplexed and intrigued — an idea that remains illusive. Infinity has taken on different identities in different fields of thought.

Infinity appears in the smallest places and in the largest. But, most of us primarily associate infinity with the very large, and as such think of it as occupying vast space. Yet a segment is composed of an infinite number of points.

Just as the ancients gazed to the heavens to experience the feeling of infinity, today with computers, advanced technologies, and the imagination we can gaze to the world of the ultra minute and experience an infinite nanouniverse.

In addition to teasing our minds and freeing our spirits, infinity is an indispensable mathematical tool used to discover such ideas as the area of a circle, limits, calculating approximations, fractals, transfinite numbers... ad infinitum!

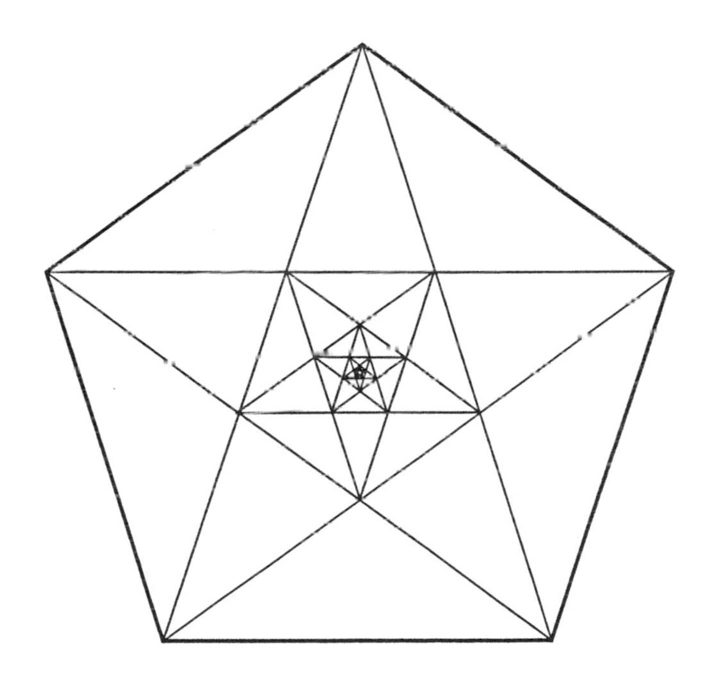

The infinite! No other question has ever moved so profoundly the spirit of man.

—David Hilbert
(1862-1943)

Oh moment, one and infinite!

—Robert Browning
(1812-1889)
By the Fireside

...infinity is not a large number or any kind of number at all; at least of the sort we think of when we say "number." It certainly isn't the largest number that could exist, for there isn't any such thing.

—Isaac Asimov
(1920-)
Asimov On Numbers

There is no smallest among the small and no largest among the large; But always something still smaller and something still larger.

— Anaxagoras
(500?-428 B.C.)

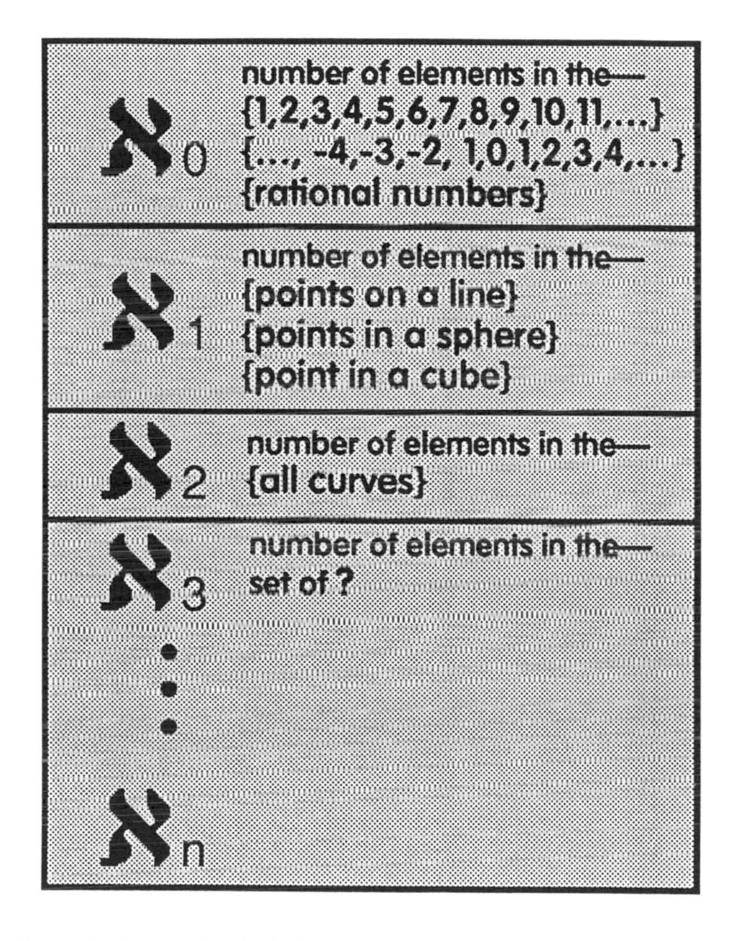

I see it, but I don't believe it!

— **Georg Cantor**
(1845-1918)

I could be bounded in a nutshell, and count myself a king of infinite space.

—William Shakespeare
(1564-1616)
Hamlet

I saw... a quantity passing through infinity and changing its sign from plus to minus. I saw exactly how it happened...but it was after dinner and I let it go."

—Sir Winston Churchill,
(1874-1965)

To see the world in a grain of sand.
And heaven in a wildflower:
Hold infinity in the palm of your hand,
And eternity in an hour.

— William Blake
(1757-1827)
Auguries of Innocence

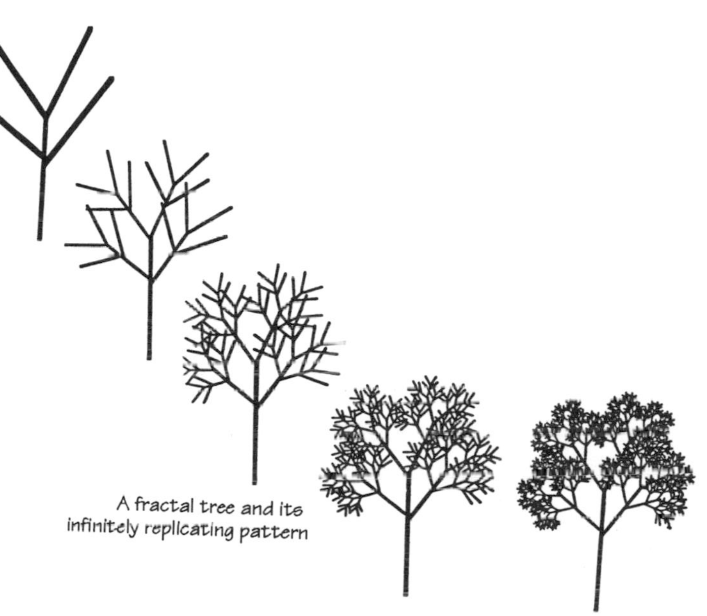

A fractal tree and its infinitely replicating pattern

Infinities and indivisibles transcend our finite understanding, the former on account of their magnitude, the latter because of their smallness; Imagine what they are when combined. ..In spite of this men cannot refrain from discussing them.

—Galileo Galilei
(1564-1642)

Dark — heaving— boundless, endless, and sublime — The image of Eternity.

— Lord Bryon
(1788-1824)
Childe Harold's Pilgrimage

Deep, deep infinity! Quietness. To dream away from the tensions of daily living; to sail over a calm sea at the prow of a ship, toward a horizon that always recedes; stare at the passing waves and listen to their monotonous soft murmur; to dream ways into unconsciousness...

— **M.C. Escher**
(1898-1972)

Let your soul stand cool and composed before a million universes.

—Walt Whitman
(1819-1892)

As lines, so loves oblique, may well
Themselves in every angle greet
But ours, so truly parallel,
Though infinite, can never meet.

— Andrew Marvell
(1621-1678)
Definition of Love

Our souls, whose faculties can comprehend
The wondrous Architecture of the world:
And measure every wand'ring planet's course,
Still climbing after knowledge infinite.

— Christopher Marlowe
(1564-1593)
Tamburlaine

Mathematics & History

The history of mathematics is not the history of concrete events, but rather the history of abstractions. It is the history of inventing ideas, of discovering patterns — it is the history of connecting these ideas and patterns to one another and to the concrete elements in our life. It is truly a history where one not only learns from the past but builds on the past — where the present relies on the past, and is essential to the future.

Here we find such things as the ancient Pythagorean theorem still useful in so many areas of mathematics. We find the mathematical monsters of the 19th century evolving to the mathematical treasure trove of fractals in the 20th century. It is where Euclid's Parallel postulate of 300 B.C. was the catalyst for discovering the non-Euclidean geometries of the 1800s. It is where each new idea is dependent on a trail of ideas from the past. Here an impossible or insignificant concept blossoms into an essential mathematical force — Eratosthenes' ancient method for sifting out prime numbers evolved to electronic locks and keys used in cryptography. One thing always leads to the next in the endless evolution of the mysterious sequence of mathematical ideas.

And so it is futile to compartmentalize mathematics by such designations as arithmetic, algebra, geometry, trigonometry, topology, calculus — each area relies on the legacies of the past and should be considered the world mathematics.

MATHEMATICS

SET THEORY · ARITHMETIC · ALGEBRA · TOPOLOGY · GEOMETRY · CALCULUS · TRIGONOMETRY · PROBABILITY · ANALYSIS · SET THEORY · SERIES · SYMBOLIC LOGIC · GAME THEORY · GROUP THEORY · STATISTICS · FUZZY LOGIC · VECTORS FRACTAL GEOMETRY · NUMBER THEORY · NON-EUCLIDEAN GEOMETRIES · COMPLEX VARIABLE · CHAOS THEORY · COMPUTER SCIENCE · KNOT THEORY · THEORY OF FUNCTIONS · THEORY OF EQUATIONS · COMPLEXITY · ...

In these days of conflicts between ancient and modern studies, there must surely be something to be said for a study which did not begin with Pythagoras and will not end with Einstein, but is the oldest and youngest of all.

— **Godfrey H. Hardy**
(1877-1947)

...a theorem as 'the square of the hypotenuse of a right angled triangle is equal to the sum of the squares of the sides' is as dazzlingly beautiful now as it was in the day when Pythagoras discovered it.

— **Charles Dodgson**
(Lewis Carroll 1832-1898)

Without the concepts, methods and results found and developed by previous generations right down to Greek antiquity one cannot understand either the aims or the achievements of mathematics in the last fifty years.

— **Hermann Weyl**
(1885-1955)

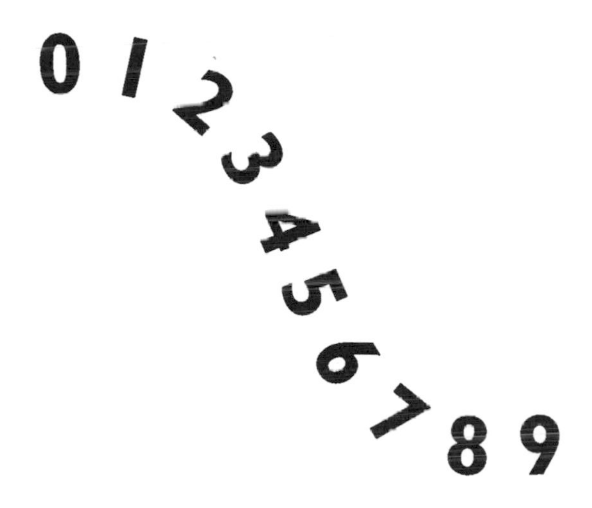

...the ingenious method of expressing all numbers by means of ten symbols, each symbol receiving a value of position as well as an absolute value... appears so simple to us now that we ignore its true merits. But its very simplicity...puts our arithmetic in the first rank of useful inventions ...remember it escaped the genius of Archimedes and Apollonius....

—Pierre-Simon Laplace
(1749-1827)

For contrary to the unreasoned opinion of the ignorant, the choice of a system of numeration is a mere matter of convention.

— Blaise Pascal
(1623-1662)

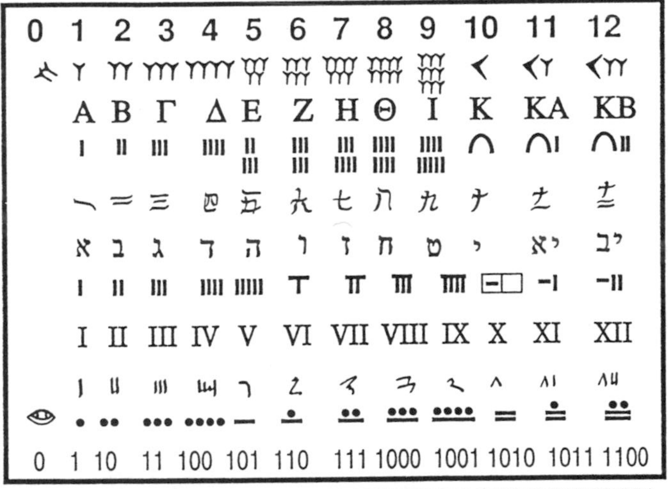

·HINDU-ARABIC·BABYLONIAN·GREEK·EGYPTIAN HIEROGLYPHIC·CHINESE SCRIPT·HEBREW ·CHINESE ROD NUMERALS·ROMAN·EGYTPIAN HIERATIC·MAYAN·BINARY NUMERALS

NUMBER SYSTEMS

If we wish to foresee the future of mathematics, the proper course is to study its history and its present state.

—Henri Poincaré
(1854-1912)

He who understands the achievements of Archimedes and Apollonius will admire less the achievements of the foremost men of later times.

— Gottfreid Wilhelm Leibniz
(1646-1716)

The golden age of mathematics – that was not the age of Euclid, it is ours.

— Cassius Jackson Keyser
(1862-1947

Someone who began to read geometry with Euclid, when he had learned the first proposition, asked Euclid, but what shall I get by learning these things? Whereupon Euclid called his slave and said give him three pence since he must make gain out of what he learns.

— **Joannes Stobaeus**
(5th century)

There are now twenty-five centuries during which mathematicians have had the practice of correcting their errors and thereby seeing their science enriched, not improvised; this gives them the right to view the future with serenity.

— **Nicholas Bourbaki**
(collective name for a predominately French group of mathematicians, since 1939)

Euclid alone
has looked on Beauty bare. Fortunate they
Who, though once only and then but far away
Have heard her massive sandal set on stone.

— **Edna St. Vincent Millay**
(1892-1950)
Euclid Alone Has Looked On Beauty Bare

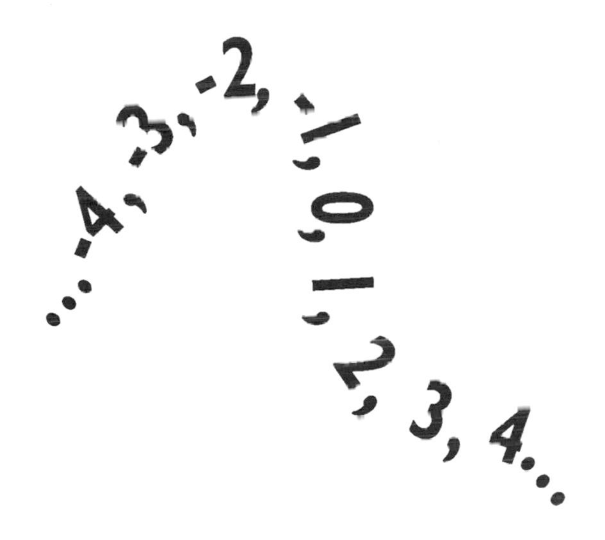

...regarding the fundamental investigations of mathematics, there is no final ending...no first beginning.

— **Felix Klein**
(1849-1925)

Mathematics & Nature

Ever wonder why a soap bubble is spherical or why a leaf can be divided exactly in half? Ever notice the shapes of certain curves recurring in the growth patterns in shells, pinecones, artichokes, the growth of human hair or the branches and bark of redwood trees? Ever in awe of land shapes or crystal formations? Nature abounds with examples of mathematical concepts. In our search to understand how things are formed, we look for patterns and similarities that can be measured and categorized. This is how mathematics is used to explain natural phenomena.

As nature puts forth its wonders, most of us are oblivious to the massive calculations and mathematical work needed to explain something that is very routine to nature. For example, the Orb spider's web is a simple, but elegant natural creation. When this beautiful structure is analyzed,

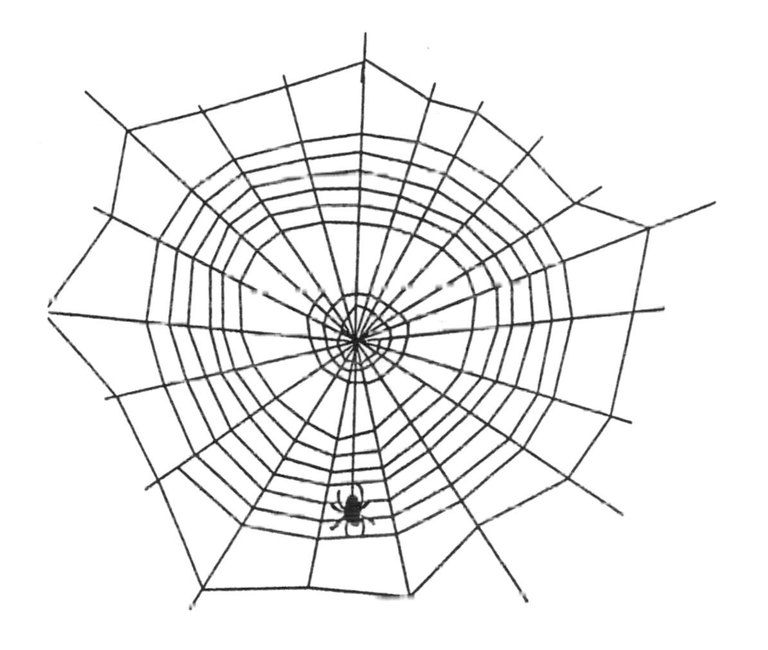

the mathematical ideas that appear in the web are indeed complicated and surprising — radii, chords, parallel segments, triangles, congruent corresponding angles, the logarithmic spiral, the catenary curve and the transcendental number e. Yet even with all our mathematical forces at work — including chaos and complexity theories — many natural phenomena, such as earthquake and weather predictions, still elude precise mathematical description.

The profound study of nature is the most fertile source of mathematical discoveries.

—Joseph Fourier
(1768-1830)

Bees ... by virtue of a certain geometrical forethought... know that the hexagon is greater than the square and the triangle and will hold more honey for the same expenditure of material.

— Pappus of Alexandria
(c. 320 A.D.)

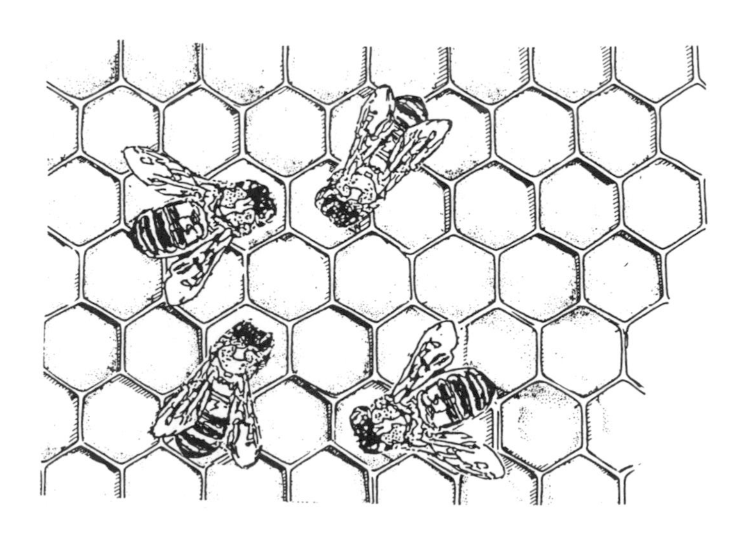

Thou, nature, art my goddess; to thy laws my services are bound...

—Karl Friedrich Gauss
(1777-1855)

For many parts of nature can neither be invented with sufficient subtlety, nor demonstrated with sufficient perspicuity, nor accommodated unto use with sufficient dexterity without the aid and intervention of mathematics.

— Galileo Galilei
(1564-1642)

Although to penetrate into the intimate mysteries of nature and hence to learn the true causes of phenomena is not allowed to us, nevertheless it can happen that a certain fictive hypothesis may suffice for explaining many phenomena.

—Leonhard Euler
(1707-1783)

...I have resolved to quit only abstract geometry,...in order to study another kind of geometry, which has for its objects the explanation of the phenomena of nature.

—René Descartes
(1596-1650)

Mathematical analysis is as extensive as nature herself.

—Joseph Fourier
(1768-1830)

Nature is pleased with simplicity, and affects not the pomp of superfluous causes.

— Issac Newton
(1642-1727)

Who by vigor of mind almost divine, the motions and figures of planets, the paths of comets, and the tides of the seas first demonstrated.

—Newton's epitaph
(1727)

Not chaos-like, together crushed and bruised,
But, as the world harmoniously confused;
Where order in variety we see,
And where, though all things differ, all agree.

— Alexander Pope
(1688-1744)

There is no branch of mathematics, however abstract, which may not some day be applied to phenomena of the real world.

— Nikolai Lobachevsky
(1793-1856)

Tiger! Tiger! burning bright
In the forests of the night,
What immortal hand or eye
Could frame thy fearful symmetry?

—William Blake
(1757-1827)
The Tiger

...the desire to understand nature has had on the development of mathematics the most important and happiest influence.

— Henri Poincaré
(1854-1912)

All the effects of nature are only mathematical results of a small number of immutable laws.

—Pierre Simon de Laplace
(1749-1827)

Our experience hitherto justifies us in believing that nature is the realization of the simplest conceivable mathematical ideas.

— Albert Einstein
(1879-1955)

Without dimension, where length,
breath, and height,
And time and place are lost; where eldest night
And chaos, ancestors of nature, hold
Eternal anarchy.

— John Milton
(1608-1674)
Paradise Lost

Everything in nature adheres to the cone, the cylinder, and the cube.

—Paul Cézanne
(1839-1906)

In Nature's infinite book of secrecy
A little I can read.

— William Shakespeare
(1564-1616)
Antony & Cleopatra

Repetition is the only form of performance that nature can achieve.

— George Santayana
(1863-1952)

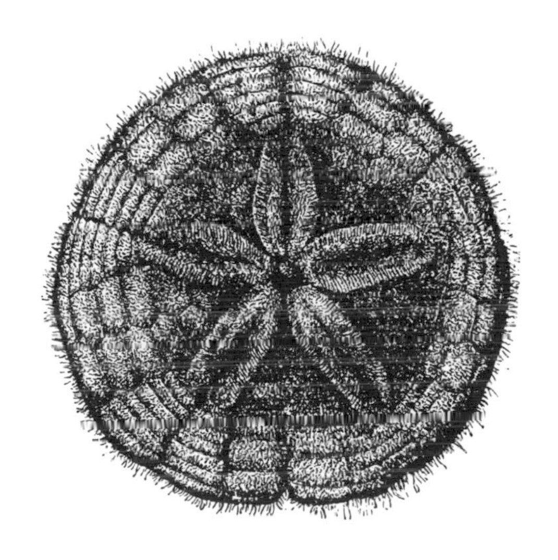

God has put a secret art into the forces of Nature so as to enable it to fashion itself out of chaos into a perfect world system.

— Immanuel Kant
(1724-1804)

Reality, Truth & Mathematics

Is it or isn't it? Real or unreal? True or False? The quest for what is real has always preoccupied the human mind. It is amazing that something dealing with imaginary things becomes an invaluable tool for describing things of reality. The focus of mathematics is to search for truths and to logically develop and prove a body of knowledge and ideas. Although these are truths about unreal objects such as numbers, rectangles, pyramids, operations, equations, it is fascinating to see how these objects fit together and function. But it is equally astonishing to discover how they function in our reality — be they hyperspace helping to unify Einstein's theories and quantum theory, the study of knots working with superstrings to explain the creation of the universe, nanotechnology and fuzzy logic to develop a realistic artificial intelligence, the theory of complexity to explain such things as economics and environmental phenomena, fractal geometry to describe nature's unusual shapes, or informational geometry and chaos theory to explore the workings of the brain and body. These are very current mathematical topics. They are ideas which scientists from all fields are using to seek truths and understand reality.

A "spiral" design? No. It is made from concentric circles, and was discovered by Dr. James Fraser in the early 1900s.

As far as the laws of mathematics refer to reality, they are not certain, and as far as they are certain, they do not refer to reality.

* * *

I want to know how God created this world. I am not interested in this or that phenomenon. I want to know His thoughts, the rest are details.

* * *

But the principle resides in mathematics. In a certain sense, I hold true that pure thought can grasp reality, as the ancients dreamed.

* * *

People like us, who believe in physics, know that the distinction between past, present, and future is only a stubbornly persistent illusion.

* * *

We are in the position of a little child entering a huge library whose walls are covered to the ceiling with books in many different tongues...The child does not understand the languages in which they are written. He notes a definite plan in the arrangement of books, a mysterious order which he does not comprehend, but only dimly suspects.

* * *

How can it be that mathematics, a product of human thought independent of experience, is so admirably adapted to the objects of reality?

* * *

Can human reason without experience discover by pure thinking properties of real things?

— **Albert Einstein**
(1879-1955)

...it would be a serious error to think that one can find certainty only in geometrical demonstrations or in the testimony of the senses.

—Augustin L. Cauchy
(1789-1857)

What is bountiful and definite and the object of knowledge is by nature prior to the indefinite and the incomprehensible and the ugly.

—Nicomachus
(c. 100 A.D.)

The chess board is the world, the pieces are the phenomena of the universe, the rules of the game are what we call the laws of Nature.

—Thomas H. Huxley
(1825-1895)

All that we see or seem
Is but a dream within a dream.

—Edgar Allan Poe
(1809-1849)
A Dream within a Dream

In order to reach the truth, it is necessary, once in one's life, to put everything in doubt—so far as possible.

— René Descartes
(1596-1650)

…the primary question was not What do we know, but How do we know it.

—Aristotle
(384-322 B.C.)

I value the discovery of a single even insignificant truth more highly than all the argumentation on the higher questions which fail to reach a truth.

— Galileo Galilei
(1564-1642)

...mathematicians are really seeking to behold the things themselves, which can be seen only with the eye of the mind.

— Plato
(427?-347? B.C.)

It is truth very certain that, when it is not in our power to determine what is true, we ought to follow what is most probable.

—René Descartes
(1596-1650)

In questions of science, the authority of a thousand is not worth the humble reasoning of a single individual.

—Galileo Galilei
(1564-1642)

I keep the subject constantly before me and wait till the first dawnings open little by little into full light.

— Isaac Newton
(1642-1727)

$$-1 = \left(\sqrt{-1}\right)^2 = \left(\sqrt{-1}\right) \cdot \left(\sqrt{-1}\right)$$

$$= \sqrt{(-1) \cdot (-1)} = \sqrt{1} = 1$$

$$\therefore \; -1 = 1\,?$$

Truth comes out of error more easily than out of confusion.

—Francis Bacon
(1561-1626)

If we do not expect the unexpected, we will never find it.

—Heraclitus
(6th century B.C.)

Everything must either be or not be, whether in the present or in the future.

—Aristotle
(384-322 B.C.)

There is nothing so easy but that it becomes difficult when you do it with reluctance.

– Terence
(190-159 B.C.)

Chance is perhaps the pseudonym of God when He did not want to sign.

—Anatole France
(1844-1924)

The truth as we see it today is this: The laws of nature do not determine uniquely the one world that actually exists.

—Hermann Weyl
(1885-1955)

No great artist ever sees things as they really are. If he did, he would cease to be an artist.

—Oscar Wilde
(1854-1900)

The known is finite, the unknown infinite; intellectually we stand on an island in the midst of an illimitable ocean of inexplicability. Our business in every generation is to reclaim a little more land.

—Thomas H. Huxley
(1825-1895)

Numbers

Some people think that the way certain numbers operate and the results that appear seem to possess a magical quality. What is amazing is how many types of numbers there are, plus all the different types of classification of numbers along with their individual characteristics. Learning about all the types of numbers can be overwhelming. But it is reassuring to learn that many of the various types of numbers that evolved over the centuries were sometimes perplexing to mathematicians themselves. The Pythagoreans did not want to destroy the mystique of the whole numbers by introducing such numbers as $\sqrt{2}$. In fact, legend has it that the Pythagoreans wanted to keep the irrational numbers secret. Blaise Pascal's statement — "I have known those who could not understand that to take four from zero there remains zero" shows even he was confused by negative numbers. Can you imagine the controversies the imaginary and complex numbers created when they entered the realm of numbers?

We Are Numbers

(A poem for Two Voices)

We are numbers.	
	Numbers.
First came one,	
	then two,
next three	
	and four not far behind.
Five,	
	six
counting	*naturals*
large	
	small,
five-hundred,	
	one-third,
gigantic,	
	minute,
a million,	
	one-millionth.

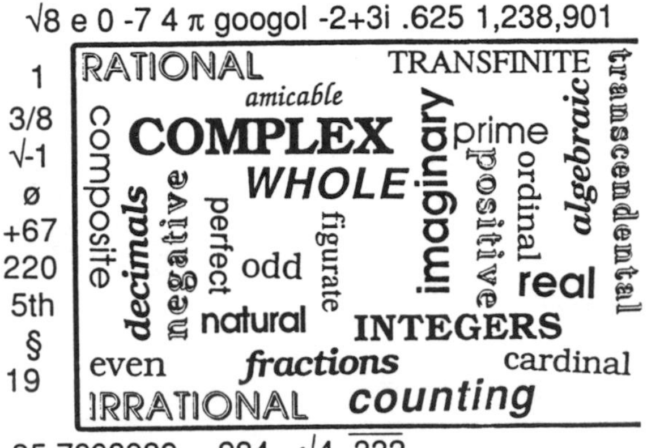

We count. *We count.*

We add and subtract.

We multiply and divide.

We keep track.

We measure.

We are numbers. *We are numbers.*

Large,

small,

never ending. *never ending.*

—Theoni Pappas
(1944-)
from *Math Talk*

No more fiction: we calculate; but that we may calculate, we had to make fiction first.

– Friedrich Nietzsche
(1844-1900)

For "is" and "is-not" though with rule and line
And "up-and-down" by logic I define,
Of all that one should care to fathom, I
Was never deep in anything but —wine.

Ah, but my computations, people say,
Reduced the year to better reckoning? —Nay,
'Twas only striking from the calendar
Unborn tomorrow and dead yesterday.

—Omar Khayyam
(1050?-1123?)
The Rubaiyat

...number is merely the product of our mind,

— **Karl Friedrich Gauss**
(1777-1855)

I have often admired the mystical way of Pythagoras, and the secret magic of numbers.

— **Sir Thomas Browne**
(1605-1682)

Six is a number perfect in itself, and not because God created all things in six days; rather the inverse is true; God created all things in six days because this number is perfect, and it would remain perfect even if the work of the six days did not exist.

— **St. Augustine**
(354-430)

3.14159265358979323846264 3
383279502884197169399375
105820974944592307816406
286208998628034825342117
067982148086513282306647
093844609550582231725359
4081284811174502841027...

π's face was marked, and it was understood that none could behold it and live. But piercing eyes looked out from the mask, inexorable, cold, and enigmatic.

— **Bertrand Russell**
(1872-1970)

{..., -3, -2, -1, 0, 1, 2, 3,...}

God made the integers, all the rest is the work of men.

— Leopold Kronecker
(1823-1891)

Quaternions came from Hamilton after his really good work had been done; and though beautifully ingenious, have been an unmixed evil to those who have touched them in any way...

— Lord Kelvin
(1824-1907)

There comes a time when an individual becomes irresistible and his action becomes all-pervasive in its effect. This comes when he reduces himself to zero.

—Mahatma Gandhi
(1869-1948)

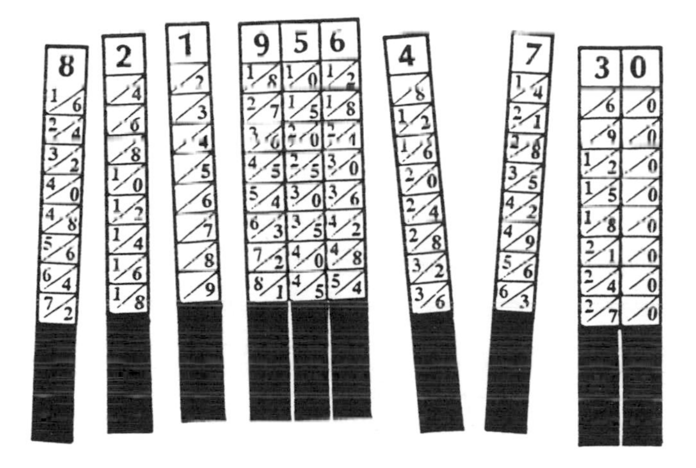

John Napier developed these rods (called Napier's bones) to help with difficult computations. They are based on logarithms which he invented in the 1590's.

There is nothing so troublesome to mathematical practice...than multiplications, divisions, square and cubical extractions of great numbers...I began therefore to consider...how I might remove those hindrances.

— John Napier
(1550-1617)

...the sole object of science is the honor of the human spirit and that under this view a problem of numbers is worth as much as a problem on the system of the world.

—Carl G.J. Jacobi
(1804-1851)

... space has a reality outside our mind whose laws cannot a priori completely prescribe.

— Karl Friedrich Gauss
(1777-1855)

Wherever there is number, there is beauty.

—Proclus
(410-485 A.D.)

... an irrational number... lies hidden in a kind of cloud of infinity.

— Michael Stifel
(1487-1567)

...it took men about five thousand years, counting from the beginning of number symbols, to think of a symbol for nothing.

—Isaac Asimov
(1920-)
Asimov On Numbers

Symbols for zero

Babylonian

Mayan

Arabic

Logic, Intuition, and Thought

To be a creative thinker one cannot rely only on logic. To be a rational thinker one cannot rely solely on intuition. But the combination of logic and intuition provide the fuel for discovery.

"Contrariwise," continued Tweedledee "if it was so, it might be; and if it were so, it would be; but as it isn't, it ain't. That's ***logic.****"*

—Lewis Carroll
(Charles Dodgson 1832-1898)
Alice In Wonderland

...mathematics presents the most brilliant example of how pure reason may successfully enlarge its domain without the aid of experience.

—Immanuel Kant
(1724-1804)

But where the senses fail us, reason must step in.

— **Galileo Galilei**
(1564-1642)

There can never be surprises in logic.

— **Ludwig Wittgenstein**
(1889-1951)

$1=2$?

If $a=b$ & $a,b>0$, then $1=2$.

Proof

1) $a, b>0$	given
2) $a=b$	given
3) $ab=b^2$	x of equals, step 2
4) $ab-a^2=b^2-a^2$	- of equals, step 3
5) $a(b-a)=(b+a)(b-a)$	factoring, step 4
6) $a=b+a$	÷ of equals, step 5
7) $a=2a$	subtitution, steps 2&6
8) $1=2$	÷ of equals, step 8

Seek simplicity, and distrust it.

— Alfred N. Whitehead
(1861-1947)

Intuition is the conception of an attentive mind, so clear, so distinct, and so effortless that we cannot doubt what we have so conceived.

— René Descartes
(1854-1912)

Everything is vague to a degree you do not realize till you have tried to make it precise.

— Bertrand Russell
(1872-1970)

Everything should be as simple as possible —
but no simpler.

— Albert Einstein
(1879-1955)

The two eyes of exact science are mathematics
and logic…

—Augustus De Morgan
(1806-1871)

As for logic, it's in the eye of the logician.

—Gloria Steinem
(1934-)
The First Ms. Reader

There is no sharply drawn line between those contradictions which occur in the daily work of every mathematician...and the major paradoxes which provide food for logical thought for decades and sometimes centuries.

—Nicholas Bourbaki
(collective name for a predominately French group of mathematicians, since 1939)

Beware gentle knight, there is no greater monster than reason.

— Cervantes
(1547-1616)

...knowledge provided by our hearts and instinct is necessarily the basis on which our reasoning has to build its conclusions.

— Blaise Pascal
(1623-1662)

A good puzzle should demand the exercise of our best wit and ingenuity, and although a knowledge of mathematics...and...of logic are often of great service in the solution of these things, yet it sometimes happens that a kind of natural cunning and sagacity is of considerable value.

—Henry E. Dudeney
(1862-1943)

Logic is the art of going wrong with confidence.

—Morris Kline
(1908-1992)

Humble thyself, impotent reason!

—Blaise Pascal
(1623-1662)

Obvious is the most dangerous word in mathematics .

—Eric Temple Bell
(1883-1960)

...it would be a serious error to think that one can find certainty only in geometrical demonstrations or in the testimony of the senses.

—Augustin Cauchy
(1789-1857)

To follow foolish precedents, and wink with both our eyes, is easier than to think.

—William Cowper
(1731-1800)

PARADOXES

Epimenides' paradox:
Epimenides was from the island of Crete and his paraodx simply states: "All Cretans are liars."

Anonymous paradox:
"Please ignore this statement."

...there is a subjective nature to life and the universe. True or false, yes or no logic cannot deal with the ever changing status in the universe. These are things that fuzzy logic deals with. Traditional logic relies on statements being either true or false. Fuzzy logic considers the degree of truth ...

— Theoni Pappas
(1944-)
The Mathematics Calendar 1995

Mathematics and the Universe

How old is the universe? How large is the universe? How did the universe begin? Is the universe growing, shrinking or unchanging? Are there such things as universal constants? What are the various shapes of paths of celestial objects? — all questions requiring mathematical answers.

Does mathematics form the framework of the universe?

Physicists, chemists, geologists, biologists and other scientists explore objects that exist in our universe. From their findings they draw hypotheses and conclusions that relate to our world. The mathematician — on the other hand — studies, explores, and creates objects that exist in their own universes, and yet their work often helps us discover ideas about the universe.

Time and again the scientific world uses mathematical ideas in its efforts to unravel the mysteries of universe. It is amazing that a subject that deals with inanimate abstract objects can give explanations for the real world.

Philosophy is written in this grand book—I mean the universe—which stands continually open to our gaze, but it cannot be understood unless one first learns to comprehend the language and interpret the characters in which it is written. It is written in the language of mathematics, and its characters are triangles, circles, and other geometric figures, without which it is humanly impossible to understand a single word of it; without these, one is wandering in a dark labyrinth.

—Galileo Galilei
(1564-1642)

The eternal mystery of the world is its comprehensibility.

—Albert Einstein
(1879-1955)

From the intrinsic evidence of his creations, the Great Architect of the Universe now begins to appear as a pure mathematician.

—James H. Jeans
(1877-1946)

We apprehend time only when we have marked motion...not only do we measure movement by time, but also time by movement because they define each other.

— Aristotle
(384-322 B.C.)

The chief aim of all investigations of the external world should be to discover the rational order and harmony which has been imposed on it by God and which He revealed to us in the language of mathematics.

—Johannes Kepler
(1571-1630)

For since the fabric of the universe is most perfect and the work of a most wise Creator, nothing at all takes place in the universe in which some rule of maximum or minimum does not appear.

—Leonhard Euler
(1707-1783)

It's ironic that fractals, many of which were invented as examples of pathological behavior, turn out not to be pathological at all. In fact they are the rule in the universe. Shapes which are not fractal are the exception. I love Euclidean geometry, but it is quite clear that it does not give a reasonable presentation of the world. Mountains are not cones, clouds are not spheres, trees are not cylinders, neither does lightning travel in a straight line. Almost everything around us is non-Euclidean.

— Benoit Mandelbrot
(1924-)

When we try to pick out anything by itself we find it hitched to everything else in the universense.

—John Muir
(1838-1914)

The progress of the human race in understanding the universe has established a small corner of order in an increasingly disordered universe.

— Stephen Hawking
(1942-)
A Brief History of Time

Joy and amazement at the beauty and grandeur of this world of which man can just form a faint notion.

— Albert Einstein
(1879-1955)

Listen there's a hell of a universe next door: let's go!

—e.e. cummings
(1894-1962)

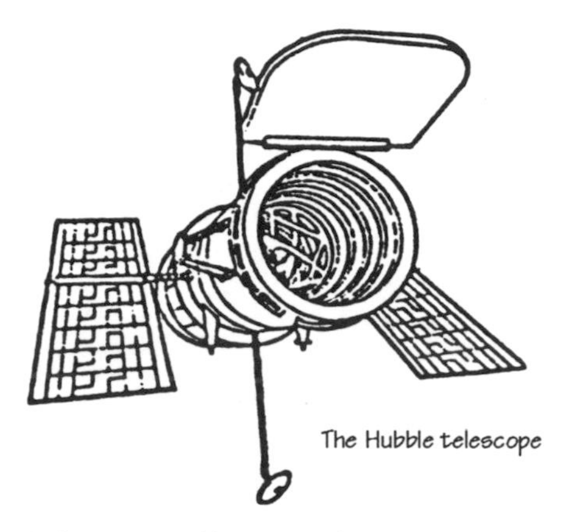
The Hubble telescope

We find them smaller and fainter, in constantly increasing numbers, and we know that we are reaching into space, farther and farther, until, with faintest nebulae that can be detected with the greatest telescopes, we arrive at the frontier of the unknown universe.

—Edwin Hubble
(1889-1953)

Where the telescope ends, the microscope begins. Who is to say of the two, which has the grandeur view?

—Victor Hugo
(1802-1885)
Les Miserables

Everything existing in the Universe is the fruit of chance and necessity.

—Democritos
5th century B.C.

As God calculates so the world is made.

—Gottfried Leibniz
(1646-1716)

...the universe was scrawled...along all its dimensions...space didn't exist and perhaps had never existed.

—Italo Calvino
(1923-1985)
A Sign in Space

The heavens themselves, the planets, and this center,
Observe degree, priority, and place,
Insisture, course, proportion, season, form,
Office, and custom, in all line of order.

—William Shakespeare
(1564-1616)
Troilus and Cressida

The world is either the effect of cause or of chance. If the latter...it is a regular and beautiful structure.

—Marcus Aurelius
(121-180 A.D.)

When they come to model Heaven
And calculate the stars, how they will wield
The mighty frame, how build, unbuild, contrive
To save appearances, how gird the sphere
With centric and eccentric scribbled o'er,
Cycle and epicycle, orb in orb.

—John Milton
(1608-1674)
Paradise Lost

We never cease to stand like curious children before the great Mystery into which we were born.

—Albert Einstein
(1879-1955)

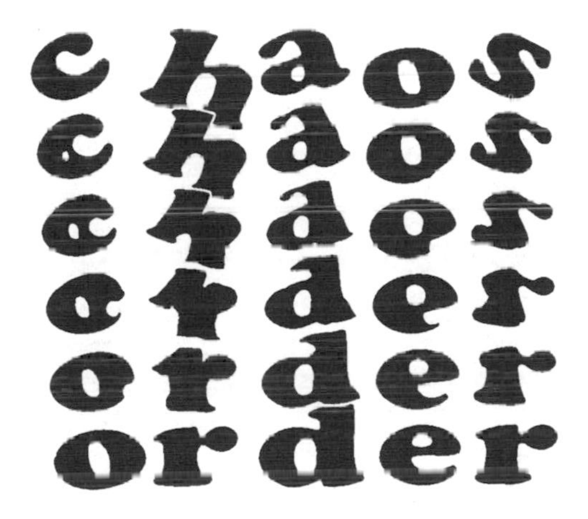

It would only be possible to imagine life or beauty as being "strictly mathematical" if we ourselves were such infinitely capable mathematicians as to be able to formulate their characteristics in mathematics so extremely complex that we have never yet invented them.

—Theodore Andrea Cook
(1867-1928)

Mathematical things

Mathematical things — numbers, statistics, fractals, cyberspace, dimensions, polyhedra, tessellations — have a pervasive way of creeping into our everyday experiences until the objects seem to become household terms. How does this happen? Through comments made by newscasters, statespeople, artists, writers, philosophers, scientists, musicians, architects, people in all areas of life. Why? Because mathematical things help to measure, describe, predict, and quantify so many facets of our lives. Be it things that deal with our bodily functions, our economics, our environment, politics — almost anything you can name will someway have mathematics connected to it. Name ten things you use or do in a day and see how many of these have something mathematical linked to them.

A marvelous neutrality have these things Mathematical, and also a strange participation between things supernatural...,and things natural...

— **John Dee**
(1527-1608)

I have discovered such wonderful things that I was amazed...out of nothing I have created a strange new universe.

—**Janos Bolyai**
(1802-1860)
from a letter to his father, 1823

The advancement and perfection of mathematics are intimately connected with the prosperity of the State.

—**Napoleon I**
(1849-1925)

I make no question but you will readily allow the square of 16 to be the most magically magical of any magic square ever made by any magician.

—Benjamin Franklin
(1706-1790)

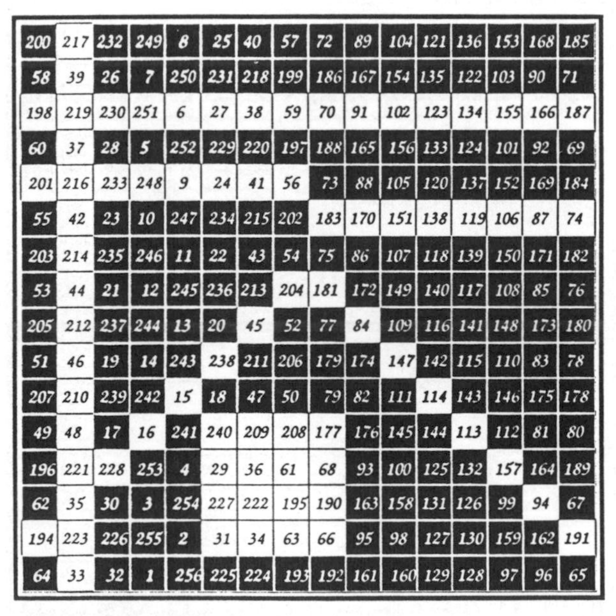

200	217	232	249	8	25	40	57	72	89	104	121	136	153	168	185
58	39	26	7	250	231	218	199	186	167	154	135	122	103	90	71
198	219	230	251	6	27	38	59	70	91	102	123	134	155	166	187
60	37	28	5	252	229	220	197	188	165	156	133	124	101	92	69
201	216	233	248	9	24	41	56	73	88	105	120	137	152	169	184
55	42	23	10	247	234	215	202	183	170	151	138	119	106	87	74
203	214	235	246	11	22	43	54	75	86	107	118	139	150	171	182
53	44	21	12	245	236	213	204	181	172	149	140	117	108	85	76
205	212	237	244	13	20	45	52	77	84	109	116	141	148	173	180
51	46	19	14	243	238	211	206	179	174	147	142	115	110	83	78
207	210	239	242	15	18	47	50	79	82	111	114	143	146	175	178
49	48	17	16	241	240	209	208	177	176	145	144	113	112	81	80
196	221	228	253	4	29	36	61	68	93	100	125	132	157	164	189
62	35	30	3	254	227	222	195	190	163	158	131	126	99	94	67
194	223	226	255	2	31	34	63	66	95	98	127	130	159	162	191
64	33	32	1	256	225	224	193	192	161	160	129	128	97	96	65

Benjamin Franklin's super duper 16x16 magic square.

I coined fractal from the Latin adjective fractus. The corresponding Latin verb frangere means 'to break': to create irregular fragments...how appropriate for our needs!

— **Benoit Mandlebrot**
(1924-)

The mathematical phenomenon always develops out of simple arithmetic, so useful in everyday life, out of numbers, those weapons of the gods; the gods are there, behind the wall, at play with numbers.

—**Le Corbusier**
(1887-1965)

The relatively simple problems—the determination of the diagonal of a square and that of the circumference of a circle—revealed the existence of new mathematical beings for which no place could be found within the rational domain.

—**Tobias Dantzig**
(1884-1956)

There are three kinds of lies: lies, damned lies, and statistics.

—Benjamin Disraeli
(1766-1848)

Statistical thinking will one day be as necessary for efficient citizenship as the ability to read and write.

— H. G. Wells
(1866-1946)

God eternally geometrizes.

—Plato
(427?-347 B.C.)

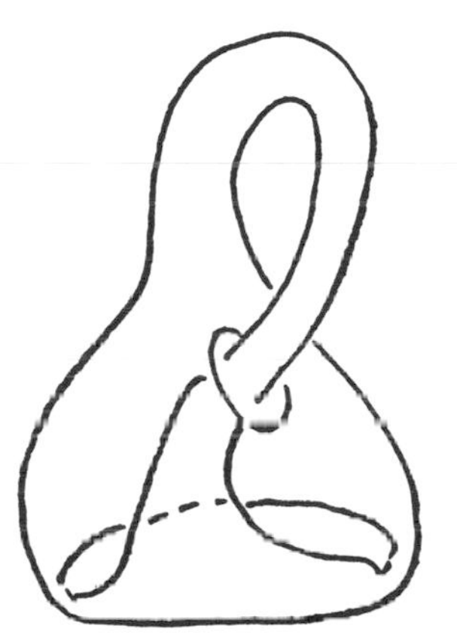

The Klein bottle passes through itself. It has a single surface with an outside but no inside.

...a mathematical subject is not considered exhausted until it has become intuitively evident...

—Felix Klein
(1849-1925)

Let...the parallel lines meet, and let me see them meet, myself — I shall see, and I shall say that they have met, but I still won't accept it.

—Fyodor Dostoyevsky
(1821-1881)
The Brothers Karamazov

The knowledge at which geometry aims is the knowledge of the eternal.

—Plato
(c. 428-348 B.C.)

Angling may be said to be so like the mathematics, that it can never be fully learnt.

—Izaak Walton
(1593-1683)
The Compleat Angler

Chance, too, which seems to rush along with slack reins, is bridled and governed by law.

—Boethius
(ca. 480-525),
The Consolation of Philosophy

... for the 5th dimension ...you can travel through space without having to go the long way around...In other words a straight line is not the shortest distance between two points.

—Madeleine L'Engle
(1941-)
A Wrinkle in Time

...the Pendulum...governed by the square root of the length of the wire and by π, that number which, however irrational to sublunar minds, through a higher rationality binds the circumference and diameter of all possible circles.

—Umberto Eco
(1932-)
Foucault's Pendulum

...Nobody knew then that there could be space. ... in reality there wasn't even space to pack us into. Every point of each of us coincided with every point of each of the others in a single point.

—Italo Cavino
(1923-1985)
All At One Point

Mathematics & the Sciences

It seems impossible to separate the sciences from mathematics because the various sciences are so dependent on mathematics for experiments and measurements. But the converse is not necessarily true. We know mathematics would continue to flourish and grow as it has in the past without relying on the sciences. Some people contend that the various applications that the sciences give to mathematical ideas help it come to life. But mathematicians know to the contrary. Math for math's sake is just fine, yet the math-science connection is crucial for unraveling scientific mysteries.

Mathematics is the gate and key to all sciences.

—Roger Bacon
(1214?-1294)

In any particular theory there is only as much real science as there is mathematics.

–Immanuel Kant
(1724-1804)

...in the future more and more theoretical physics will command a deep knowledge of mathematical principles; and also that mathematicians will no longer limit themselves so exclusively to the aesthetic development of mathematical abstractions.

—George David Birkoff
(1884-1944)

All science as it grows toward perfection becomes mathematical in its ideas.

— Alfred North Whitehead
(1861-1947)

Mechanics is the paradise of mathematical science because here we come to the fruits of mathematics.

— Leonardo da Vinci
(1452-1519)

As long as a branch of science offers an abundance of problems, so long is it alive.

— David Hilbert
(1862-1943)

In most sciences one generation tears down what another has built and what one has established another undoes. In mathematics alone each generation adds a new story to the old structure.

—Hermann Hankel
(1839-1873)

It is the perennial youthfulness of mathematics itself which marks it off with a disconcerting immortality from the other sciences.

—Eric Temple Bell
(1883-1960)

Science has explored the microcosmos and the macrocosmos...The great unexplored frontier is complexity...I am convinced that nations and people that master the new science of Complexity will become the economic, cultural, and political superpowers of the next century.

—Heinz Pagels
(1939-1988)

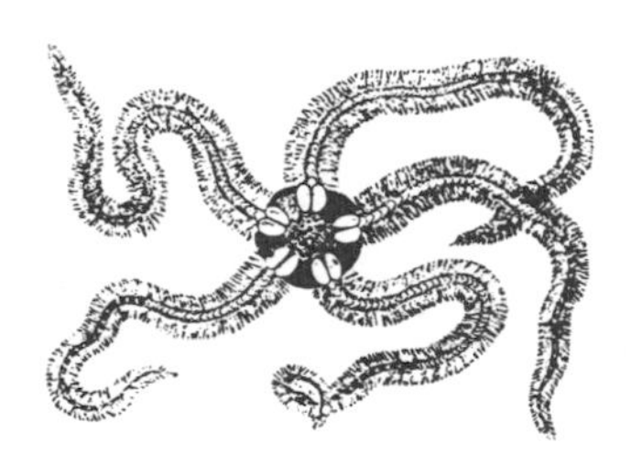

...no human inquiry can be called science unless it pursues its path through mathematical exposition and demonstration.

– Leonardo da Vinci
(1452-1519)

In science, each new point of view calls forth a revolution in nomenclature.

—Friedrich Engels
(1820-1895)

What science can there be more noble, more excellent, more useful ... than mathematics.

—Benjamin Franklin
(1706-1790)

Science is the great antidote to the poison of enthusiasm and superstition.

— Adam Smith
(1723-1790)

...All the sciences which have their end investigations concerning order and measure are related to mathematics...it far surpasses in facility and importance the sciences which depend upon it ...it embraces at once all the objects to which these are devoted and a great many others besides...

—René Descartes
(1596-1650)

Etching by Albrecht Dürer (1471-1528)

Computers

The computer is different things to different people.

To the writer it is a pen or typewriter. To the accountant it is a sophisticated calculator. For the artist —a paintbrush and palette, for the scientist a laboratory, for the architect a drafting table, for the engineer design tools, for the professor a research tool, for the librarian a card catalog and reference desk. For the mathematician the computer is a new tool which enables the testing of theories, ideas, and exploring new number frontiers, not possible without the aid of these super number crunchers.

We have heard of the term user friendly, yet many people are still uncomfortable around a computer. Even people who use computers on a daily basis at their work are reticent to venture beyond the daily routines they have mastered. Why?

Perhaps because it is mechanical. Perhaps they feel they may damage it.

Regardless of the reticence of so much of the

general populace to venture into the computer's world with full force, the computer is entering our lives at breakneck speeds— impacting almost all aspects of daily routines.

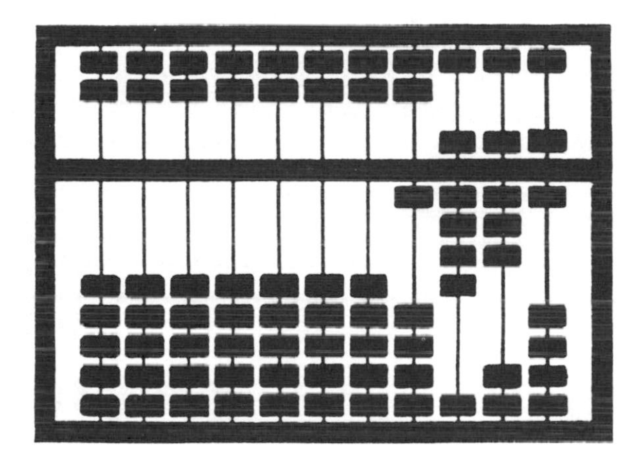

Men have become fools of their tools.

— Henry Thoreau
(1817-1862)

To err is human, but to really foul things up requires a computer.

— Anonymous

The real question is not whether machines think, but whether men do.

—B. F. Skinner
(1904-1990)

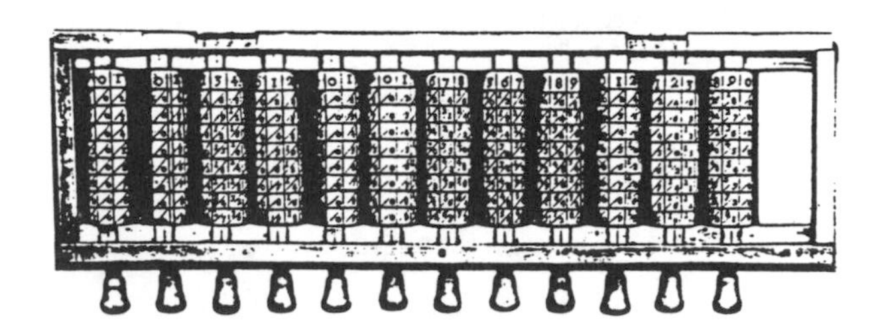

Rendition of a calculator based on Napier's rods, 17th century.

It is unworthy of excellent men to lose hours like slaves in the labor of calculation which could safely be relegated to anyone else if machines were used.

—Gottfried Wilhelm von Leibnitz
(1646-1716)

Either mathematics is too big for the human mind or the human mind is more than a machine.

—Kurt Gödel
(1906-1978)

It is not a bad definition of man to describe him as a tool-making animal.

— Charles Babbage
(?1791-1871)

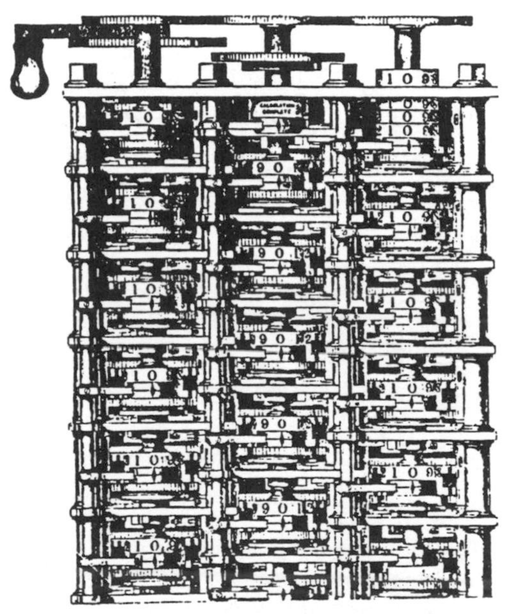

Rendition of a portion of
Charles Babbage's difference engine, 1823.

Who can foresee the consequences of such an invention.

* * *

For the machine is not a thinking being, but simply an automaton which acts according to the laws imposed upon it.

— **Ada Lovelace**
(1815-1852)

Sometimes I consider myself a fisherman. Computer programs and ideas are the hooks, rods and reels. Computer pictures are the trophies and delicious meals. ...Often the specific catch is a surprise, and this is the enjoyment of the sport. There are no guarantees. There are often unexpected pleasures.

— Clifford Pickover
(1957-)
Computers, Pattern, Chaos and Beauty

We may hope that machines will eventually compete with men in all purely intellectual fields.

— Alan Turing
(1912-1954)

Man has within a single generation found himself sharing the world with a strange new species: the computer...

— Marvin Minsky
(1927-)

We protect our computers. We bolt them to the desk and so forth...the computer right now is still more noticeable by its presence than its absence. When you go somewhere and somebody doesn't have a computer on them and that becomes a remarkable thing...then I think the computer will have made it. Its destiny is to disappear into our lives like all of the things that we don't think of as technology like wrist-watches, paper and pencils and clothing.

—Alan Kay
(1940-)

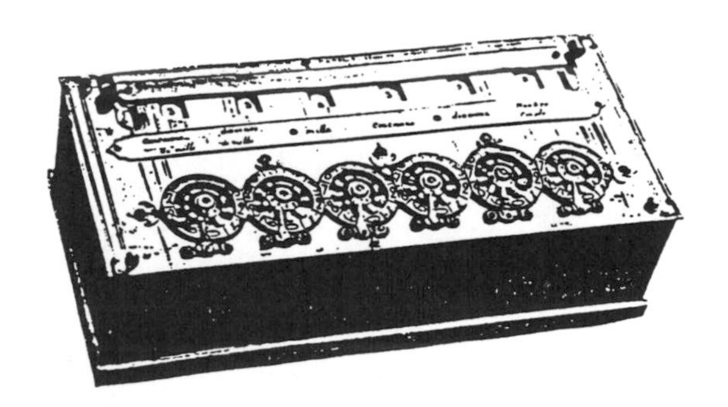

Rendition of Pascal's calculator, 1642.

When I saw people that could never possibly design a computer,...build a hardware kit,...assemble their own keyboards and monitors,...write their own software, using these things, then you knew something very big was going to happen.

— Steve Jobs
(1955-)

...It was like a revolution that I'd never seen. You read about technological revolutions, the Industrial Revolution and here was one of those sort of things happening and I was part of it.

— Stephen Wozniak
(1950?-)

Epilogue

Everything has been said before by someone.
— Alfred North Whitehead
(1861-1947)

From my early teaching years, quotations decorated my classroom walls. It always fascinated my students and me that comments made centuries ago were so relevant today.

Besides being amusing and thought provoking, quotations can shed light on a subject often avoided in everyday discussions. I have found that quotations can offer insights, inspiration and connect us more intimately to mathematics.

Mathematics teacher and consultant Theoni Pappas received her B.A. from the University of California at Berkeley in 1966 and her M.A. from Stanford University in 1967. Pappas is committed to demystifying mathematics and to helping eliminate the elitism and fear often associated with it.

In addition to **The Music of Reason** her other innovative creations include **The Math-T-Shirt, The Mathematics Calendar, The Children's Mathematics Calendar, The Mathematics Engagement Calendar,** and **What Do You See?**—an optical illusion slide show with text. Pappas is also the author of the following books: **Mathematics Appreciation, The Joy of Mathematics, Greek Cooking for Everyone, Math Talk, More Joy of Mathematics, Fractals, Googols & Other Mathematical Tales,** and **The Magic of Mathematics.**

Mathematics Titles by Theoni Pappas

THE MAGIC OF MATHEMATICS
$10.95 • 336 pages
illustrated • ISBN:0-933174-99-3

FRACTALS, GOOGOLS & Other Mathematical Tales
$9.95 • 64 pages
illustrated • ISBN:0-933174-89-6

THE JOY OF MATHEMATICS
$10.95 • 256 pages
illustrated • ISBN:0-933174-65-9

MORE JOY OF MATHEMATICS
$10.95 • 306 pages
cross indexed with *The Joy of Mathematics*
illustrated • ISBN:0-933174-73-X

MATHEMATICS APPRECIATION
$10.95 • 156 pages
illustrated • ISBN:0-933174-28-4

MATH TALK
mathematical ideas in poems for two voices
$8.95 • 72 pages
illustrated • ISBN:0-933174-74-8

THE MATHEMATICS CALENDAR
$9.95 • 32 pages • written annually
illustrated • ISBN:1-884550-

THE CHILDREN'S MATHEMATICS CALENDAR
$9.95 • 32pages • written annually
illustrated • ISBN:1-884550-

WHAT DO YOU SEE?
An Optical Illusion Slide Show with Text
$27.95 • 40 slides • 32 pages
illustrated • ISBN:0-933174-78-0

THE MUSIC OF REASON
$9.95 • 128 pages
illustrated • ISBN:1-884550-04-5